Contents

Acknowledgements..................
Introduction & Scope...............
Why Study Spiders?......................
Classification & Anatomy..14
Enemies of Spiders..21
Spider families
- Araneidae..22
- Cithaeronidae..26
- Corinnidae..27
- Ctenidae..28
- Deinopidae..29
- Dictynidae..30
- Eresidae..31
- Filistatidae..32
- Gnaphosidae..33
- Hahniidae..34
- Hersiliidae..35
- Linyphiidae..36
- Lycosidae..37
- Miturgidae..39
- Mysmenidae..40
- Nephilidae..41
- Ochyroceratidae..43
- Oecobiidae..44
- Oonopidae..45
- Oxyopidae..46
- Palpimanidae..48

- Philodromidae 49
- Pholcidae 50
- Pisauridae 51
- Prodidomidae 53
- Salticidae 54
- Scytodidae 58
- Segestriidae 59
- Selenopidae 60
- Sparassidae 61
- Tetragnathidae 62
- Theridiidae 63
- Thomisidae 65
- Uloboridae 68
- Zodariidae 69

Other Arachnids 70
- Solifugae 71
- Amblypygi 72
- Thelyphonida & Schizomida 73
- Scorpiones 74
- Pseudoscorpiones 75
- Acari 76
- Ricinulei 77
- Opiliones 78

Further Reading 79
Index To Genera Cited 80

Acknowledgements

The author is particularly grateful to the following in The Gambia for their hospitality, advice and discussion: all staff of Bijilo Forest, Abuko Nature Reserve, Tanji Bird Reserve and the Department of Forestry, especially Jato Sillah (director of the DoF), Lamin Bojang, Kebba Sonko, Lamin Jammeh and Sulayman Jobe (manager of Bijilo Forest), Alpha Jallow (director of the Department of Parks and Wildlife Management), Clive Barlow, Jon Baker and Luc 'Nouroudine' Paziaud (Gambian Reptiles Farm).

The following professional taxonomists are thanked for assistance with provisional identification and information: Wanda Wesolowska (Poland), Bernhard Huber (Germany), Dmitri Logunov (UK), Diana Silva Dávila (Peru), Tamas Szuts (Denmark), Patrick Blandin (France), Matjaz Kuntner (Slovenia), Galina Azarkina and Yuri Marusik (Russia), Charles Haddad, Elizabeth Kassimatis, Leon Lotz and Ansie Dippenaar-Schoeman (South Africa), Rudy Jocqué (Belgium) (spiders); Lorenzo Prendini, (USA), Frantisek Kovařík (Czech Republic) (scorpions); Alan Walker (UK, Acari); Alexander Gromov (Kazakhstan, Solifugae), Mark Judson (France), Mark Harvey (Australia) (Pseudoscorpions); Gerald Legg (UK, Ricinulei).

I am extremely grateful to Dmitri Logunov and Phil Rispin (Manchester Museum, UK) for access to, and assistance with, their extensive library collection during the preparation of this book. Barry Burke (UK) is thanked for assistance with software and Jon Baker is thanked for proof reading. I am extremely grateful to my mother for supporting me in times of difficulty and to my beautiful daughter Siri for being there!

Photo credits:
(T = top, M = middle, B = bottom, L = left, R = right)

Jon Baker (UK/The Gambia): p. 42 (BL)
Stefano Converti (Italy/The Gambia): p. 24 (ML), 28 (BR), 29, 57 (BL), 61 (TL/R), 72 (BL), 73, 74 (TR/BL)
Yuri Marusik (Russia): p 34, 40, 43, 45, 75
All other photographs by the author

Any errors are the responsibility of the author. The author is a professional zoologist who has spent several years in The Gambia. He is a Visiting Research Fellow in the Faculty of Life Sciences at The University of Manchester, UK and an internationally recognized expert on spiders. He can be contacted by email on david.penney@manchester.ac.uk or siri.press@live.co.uk and would be grateful to receive any nice Gambian wildlife photographs, particularly of spiders...many thanks!

INTRODUCTION & SCOPE

The Gambia is located on the Atlantic coast of tropical West Africa between latitudes 13 and 14 degrees north, with an open coastline of 70 km and is otherwise surrounded by Senegal for its entire land border of 740 km. There is a sheltered coastline of 200 km along the River Gambia dominated by extensive mangrove systems and mudflats. The Gambia loosely tracks the course of the river as it meanders inland (the river originates in Guinea and flows from east to west) and with an area of only 11,295 km^2 (10,000 km^2 of dry land) is the smallest country on the continent. In contrast to many other West African countries it is relatively flat, with the highest point less than 100 m above sea level. The country is situated on the Continental Terminal, a vast Tertiary sandstone plateau, which originated from the iron-rich soils to the east. These were eroded from the continent into the Atlantic and then over millions of years redeposited against the coastline and lithified (turned to stone).

The sub-tropical climate is pleasant, with two distinct seasons determined by the imbalance at the boundary of the high-pressure regions north and south of the equator (the ITCZ: inter-tropical convergence zone). Generally speaking, November to June consist of dry savannah winds (Harmattan), whereas from July to October the country sees heavy downpours and is lusciously green. Even in the height of the rainy season it does not rain every day and much of the rainfall occurs overnight. There are still many days of uninterrupted sunshine, but the high humidity can be rather oppressive towards the end of the rainy season. However, Gambian biodiversity is at its most spectacular during the rains, not least because they initiate an explosion of a great diversity of flowering plants and a visit at this time of year is highly recommended.

The population of the country is nearly 1.7 million, increasing annually with a growth rate of approximately 2.7%. Given that the major economic activities within The Gambia, such as agriculture, agro-processing, fisheries, livestock

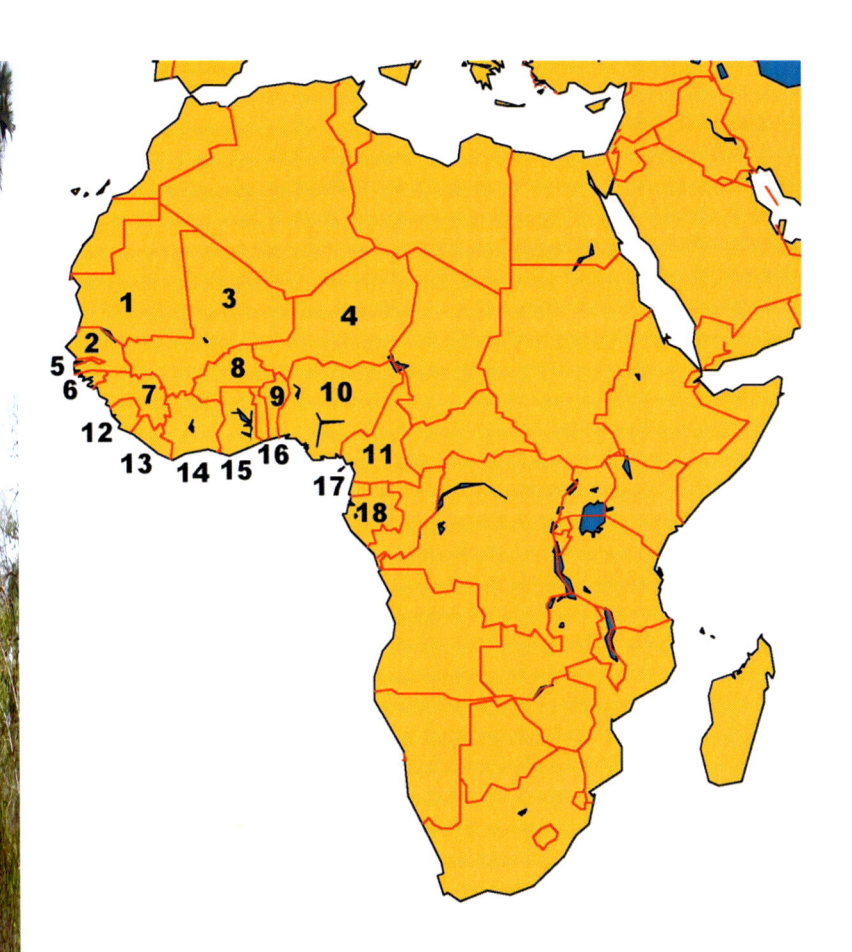

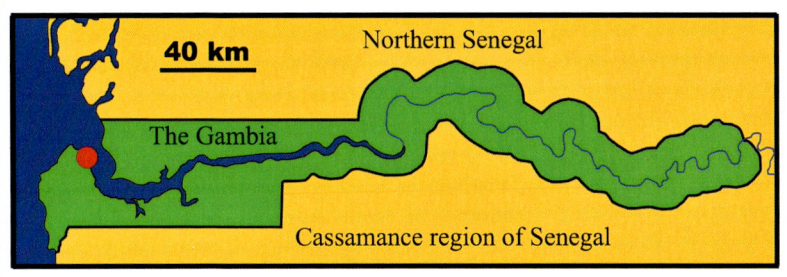

The Gambia and other tropical West African countries: **1**, Mauritania; **2**, Senegal; **3**, Mali; **4**, Niger; **5, The Gambia**; **6**, Guinea-Bissau; **7**, Guinea; **8**, Burkina Faso; **9**, Benin; **10**, Nigeria; **11**, Cameroon; **12**, Sierra Leone; **13**, Liberia; **14**, Ivory Coast; **15**, Ghana; **16**, Togo; **17**, Equatorial Guinea; **18**, Gabon. Red dot in the bottom figure represents Banjul, the capital of The Gambia.

production and tourism have quite specific land usage requirements, the increasing population size and the subsequent increase in demand for these products will undoubtedly affect the relative abundance and distribution of the various habitat types present in the country. What the knock-on effects of this will be for Gambian wildlife is unclear.

The Gambia lies within the transition zone at the interface of relatively moist Guinean forest–savannah mosaic in the south and the drier Sudanian woodlands in the north. Within this system, the country supports many diverse habitat types including marine and coastal, estuary and mangrove, banto faros (barren hypersaline flats derived from the mangroves), brackish and freshwater river banks, swamp forest, freshwater swamps and other wetlands (e.g., bolons, rice fields, ponds, etc.; obviously more prominent in the rainy season), forest (primary coastal e.g., Bijilo, primary gallery e.g., Abuko, secondary e.g., Tanji, etc.) and woodland, forest–savannah mosaic, villages, farmland and fallow land.

Such a diverse range of habitat types can be expected to contain a high biodiversity of indigenous plants and animals, in addition to naturalized introductions and migrant species. This is certainly the case, but our knowledge of the fauna and flora is far from complete and differs significantly for different groups. In general, those groups that are useful (e.g., for building materials, have medicinal properties, are edible, are dangerous to humans or their crops) tend to be better known than less so-called useful groups.

Spiders are familiar to us all as a result of their high global species biodiversity and their ubiquitous and sometimes synanthropic (living alongside humans) habitat preferences. Indeed, they are imprinted into our psyche from a very early age, through nursery rhymes and folklore. This is also true of West Africa where Anansi the spider forms a very important part of Ghanaian folk tales.

Most people have stopped, at one time or another, to marvel at the magnificent feat of engineering that is a freshly woven spider's web. Others have arachnophobia to such a degree that they cannot even bear to hear the word 'spider'. However, love them or loathe them, most people are unaware that spiders have a long geological history and that they were among the first animals to conquer the land almost 400 million years ago. We know this from studying their fossil record.

Spiders belong to the order Araneae and are arachnids not insects. They differ in having eight legs rather than six, no wings or antennae, only two parts to the body, and they possess silk-spinning organs called spinnerets. Other extant (alive today) arachnids include scorpions (order Scorpiones), pseudoscorpions (order Pseudoscorpiones), camel spiders or wind scorpions (order Solifugae), vinegaroons or tailed whip scorpions (order Thelyphonida [=Uropygi]), tailless whip scorpions (order Amblypygi), harvestmen (order Opiliones), mites and ticks (order Acari), hooded tick spiders (order Ricinulei), palpigrades (order Palpigradi) and schizomids (order Schizomida). All are known from the fossil record and all but the last two have been recorded from The Gambia (some for the first time in this book). In addition, there are a number of strictly fossil (extinct) Palaeozoic orders, namely, Uraraneida, Phalangiotarbida, Trigonotarbida and Haptopoda, none of which are known from Africa. Note that all these orders share the same basic arachnid body plan of four pairs of legs and one pair of pedipalps, but the relationships between these orders are still unresolved at their deepest levels.

The arachnids of The Gambia and West Africa in general are poorly known, yet they are extremely interesting from a biogeographical perspective. For instance, two arachnid orders are known nowhere else on the African continent. Hooded tick spiders (Ricinulei) have very localized distributions in various West African countries, with the newly discovered Gambian specimens representing the northernmost records. More restricted still, is the sole African representative of the vinegaroons or tailed whipscorpions (Thelyphonida), which has been recorded only from The Gambia and Senegal. Also

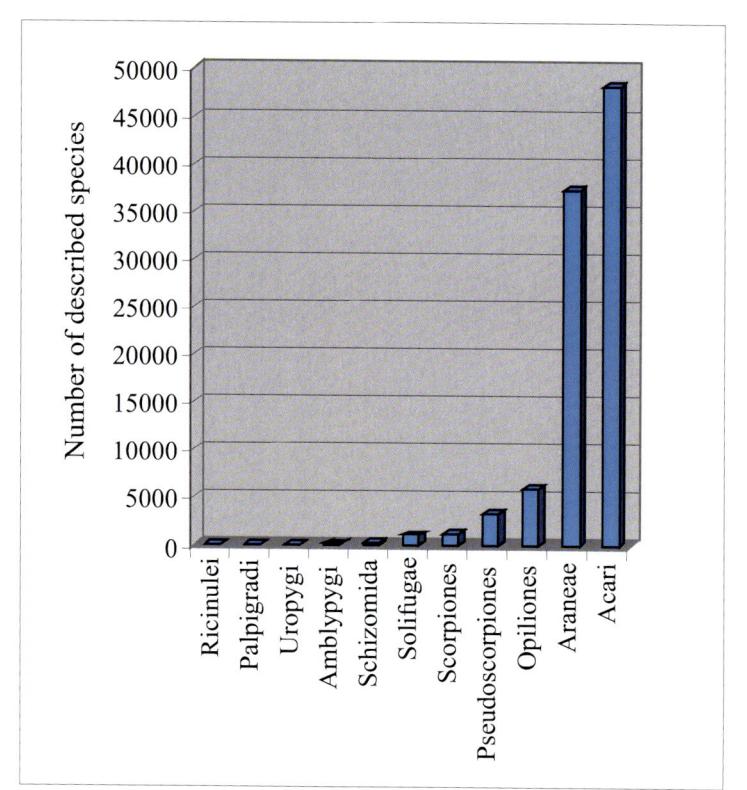

Global diversity of the arachnid orders. *Data in the above chart are taken from Harvey (2002, 2007). They include 97,920 arachnid species, but the correct value will almost certainly exceed 100,000 by now given that spiders are now known from more than 40,000 described species.*

Harvey, M.S. 2002. The neglected cousins: what do we know about the smaller arachnid orders. J. Arachnol. 30: 357–372.
Harvey, M.S. 2007. The smaller arachnid orders: diversity, descriptions and distributions from Linnaeus to the present (1758 to 2007). Zootaxa 1668: 363–380.

of interest is the primitive tailless whipscorpion suborder Paleoamblypygi, which is only known to exist in Guinea-Bissau, where it lives in termite nests. Thus, its presence in The Gambia would not be surprising. The specialist habitat requirements of many of the West African arachnid species is cause for conservation concern, particularly in a region of the world where land usage requirements are changing rapidly, usually to the detriment of natural habitats.

More than 40,000 different species of spiders in 109 families have been described to date, with more than 5,400 species in more than 70 of these families known from Africa. The excellent *African Spiders: An Identification Manual* published in 1997 will certainly form the baseline of any future studies on the spiders of Africa. It is clear from the aforementioned work that our knowledge of West African arachnid biodiversity and that of The Gambia in particular is extremely poor and there is no contemporary work dedicated to the West African arachnid fauna. The *African Arachnid Database* (AFRAD 2007) lists only five spider species for The Gambia (Corinnidae: *Castianeira loricifera*, *Trachelas punctatus*; Dictynidae: *Mizaga* sp.; Pisauridae: *Afropisaura valida*; Salticidae: *Stenaelurillus nigricaudus*) and only 141 species for Senegal. A *Provisional Checklist of all Species Recorded Within The Gambia*, published online by *Makasutu Wildlife Trust* in 2005 (updated 2006: www.darwingambia.gm) lists only three species of spiders, only one of which I can confirm to be present in The Gambia (Nephilidae: *Nephila senegalensis*), three mites (*Mononychellus tanajoa, Panonychus citri, Tetranychus cinabarinus*) and one scorpion (Buthidae: *Buthus occitanus*), with the remaining arachnid orders either omitted or listed as unrecorded.

More recently, three short research notes by the author have added several more spider species to the list (Araneidae: *Cyclosa* sp.; Salticidae: *Menemerus bivittatus*, *Holcolaetis vellerea*; Sparassidae: *Heteropoda* sp.) and even a new order (Ricinulei). A *Field Guide to Wildlife of The Gambia*, also by the author includes several additional families and species, but nothing arachnological that is not included in this work. One aim of this book is to rectify the deficit of information available to those interested in this remakable group of creatures.

During my time in The Gambia, several locals have asked me why I look at spiders, to which I normally responded by asking them whether they thought spiders were important and if they did anything useful. Invariably I was told they were of no consequence whatsoever. I then asked them if mosquitoes (which spread malaria and untold other diseases), cockroaches and flies (which foul fruits, foodstuffs and cause myiasis and sickness in livestock and people), caterpillars (that damage crops), weevils and other beetles (which form pests of stored products) and termites (that destroy wooden structures) were important. They all appreciated the importance of these groups, but most were unaware that spiders (and to an extent other arachnids) functioned as major control factors for these deleterious insects by eating them. Indeed, most seemed quite amazed by this revelation and they usually promised to stop killing them on sight, which tends to be the normal response to seeing a spider. Out of additional interest, I usually asked them if they knew how many different kinds of spider existed in The Gambia. The normal answer was around about three! It was also interesting to note the common belief that if a spider (specifically the giant golden orb-weaving spiders *Nephila* spp.) 'urinated' upon your skin then it was going to blister, become infected and cause serious medical problems.

The correct identification of many arachnid species requires detailed microscopic examination by relevant experts, which may not be available for some groups. Hence, you will notice varying degrees of taxonomic resolution. Many of the species illustrated are first records for The Gambia and some may even be new to science. This guide to Gambian arachnids is by no means exhaustive. There are certainly many other species present in the country that are not included. Indeed, there are several spider families present (or expected to be present) that are not included, primarily because I have not been able to confirm them. These include some of the mygalomorph families and I am quite certain (but cannot be one hundred percent sure) that I once saw a caponiid spider race away from under a stone! The original intention was to collect additional data and resolve further the identification of some of the species. However, significant changes in my personal circumstances prevented this.

WHY STUDY SPIDERS?

Research based on extant spiders (other than basic taxonomy, systematics, biodiversity etc.) is focussed on a variety of areas. The physical properties of spider silk include great strength and elasticity. Therefore, it is a potentially valuable substance for producing super-strong materials, and attempts are currently underway to biosynthesize and mass produce it in the laboratory. Medical applications include research into the use of spider silk as a relatively inert scaffold material for use with bioimplants, and the use of spider venom components for creating drugs to treat patients with cardiac problems. Additional venom research includes the development of organic insecticides as well as antivenoms.

In contrast to many other species rich invertebrate groups, such as mites, nematodes and various insect orders, spiders are relatively large animals. Many of these other groups require complicated, time-consuming dissection in order to identify them to species level. However, species identification of spiders is based on their species-specific, external copulatory organs, which are often clearly visible without dissection or the use of complicated techniques. Today, spiders form ideal model organisms for use in rapid biodiversity assessments because of their relative ease of collection and identification, and because of their fine-tuned distributions and habitat specificities. Indeed, some scientists believe that spiders provide information about the intrinsic biological value of any particular habitat better than higher plants or vertebrates.

From a purely biological perspective, spiders are extremely interesting in terms of their physiology and behaviour in addition to their high global biodiversity. Jumping spiders (family Salticidae) have some of the best visual acuity of all animals and also exhibit complex behaviours indicative of cognitive reasoning. Some are highly specialized. For example, one African jumping spider feeds preferentially on human blood-fed mosquitoes!

CLASSIFICATION & ANATOMY

In order to facilitate the study of the great biodiversity of living organisms, a hierarchical classification scheme, known as the Linnean system has been adopted. The formal zoological classification of spiders is as follows:

Kingdom: **Animalia**
  Phylum: **Arthropoda**
    Infraphylum: **Cheliceriformes**
      Subphylum: **Arachnomorpha**
        Superclass: **Chelicerata**
          Epiclass: **Euchelicerata**
            Class: **Arachnida**
              Subclass: **Tetrapulmonata**
                Order: **Araneae**
                  Suborder:

1. **Mesothelae** (primitive spiders with a segmented abdomen, restricted to south-east Asia)
2. **Opisthothelae**
    Infraorder:
      1. **Mygalomorphae** (tarantula-like, baboon, trap-door spiders)
      2. **Araneomorphae** (common spiders)

Within each infraorder, the following taxonomic ranks (easily identified by the suffix) are used in zoological nomenclature to further classify spiders:

Superfamily (-oidea): e.g., Araneoidea
  Family (-idae): e.g., Nephilidae
    Subfamily (-inae): e.g., Nephilinae
      Tribe (-ini): e.g., Nephilini (a rarely used level)
        Genus: e.g., *Nephila*
          Species: e.g., *Nephila senegalensis*

The basic body plan of spiders is shown below, with more specific features highlighted in the family pages. All the other arachnid orders share the same plan of two main body regions (sometimes fused and appear as one), four pairs of walking legs (the first of which can be long and antenniform to serve a sensory function (e.g., whip scorpions) and a pair of pedipalps, often highly modified to form pincers or other forms of clasping appendages. In camel spiders the pedipalps are elongated and leg-like and give the impression of the animal having ten legs.

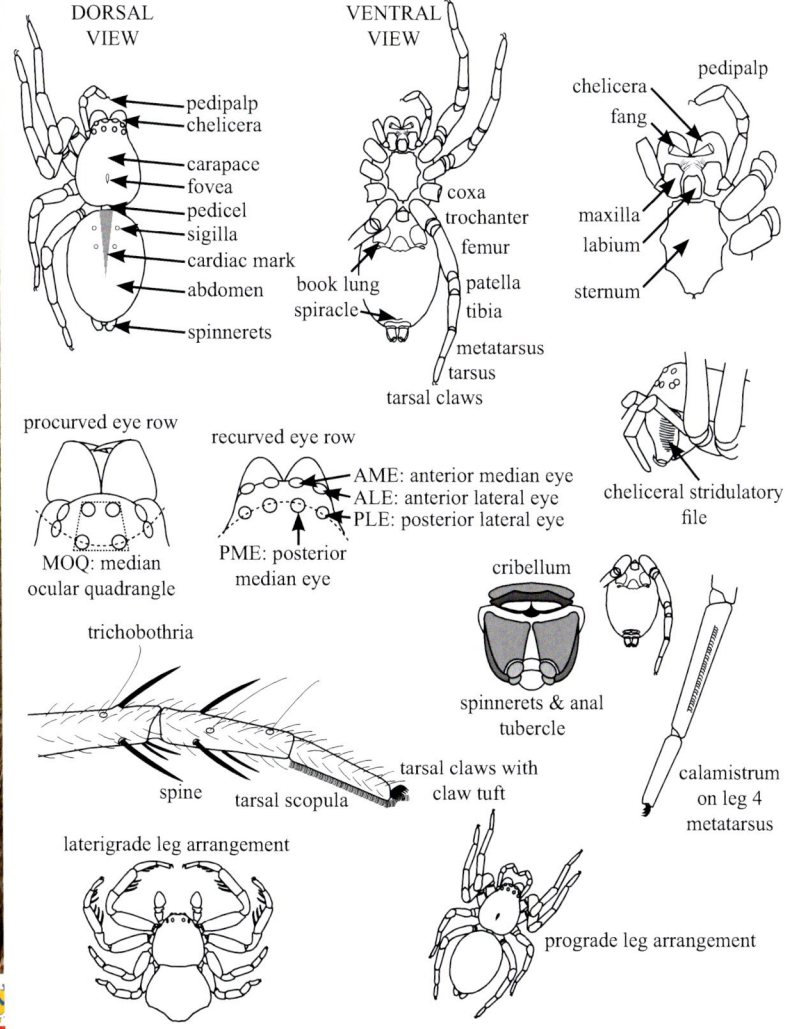

DORSAL VIEW
- pedipalp
- chelicera
- carapace
- fovea
- pedicel
- sigilla
- cardiac mark
- abdomen
- spinnerets

VENTRAL VIEW
- coxa
- trochanter
- femur
- patella
- tibia
- metatarsus
- tarsus
- tarsal claws
- book lung
- spiracle

- pedipalp
- chelicera
- fang
- maxilla
- labium
- sternum

procurved eye row

MOQ: median ocular quadrangle

recurved eye row
- AME: anterior median eye
- ALE: anterior lateral eye
- PLE: posterior lateral eye
- PME: posterior median eye

cheliceral stridulatory file

cribellum

spinnerets & anal tubercle

trichobothria

spine   tarsal scopula   tarsal claws with claw tuft

calamistrum on leg 4 metatarsus

laterigrade leg arrangement

prograde leg arrangement

15

They have two major tagma (body parts) joined by a narrow pedicel through which the circulation and feeding systems are canalized: the anterior prosoma and the posterior opisthosoma. The prosoma is covered dorsally by the carapace, which is divided into two regions: the anterior cephalic region (where the eyes are situated) and the posterior thoracic region (often containing a fovea—a distinct depression which serves as an internal point for muscle attachment). In some species the division of these regions is distinct, demarcated by a cervical groove, whereas in others it is not noticeable; both situations may occur in a single family. The shape and surface texture of the carapace (including pitting and arrangement of setae, etc.) are important taxonomic features in some cases. For example, the spider family Scytodidae (spitting spiders) are recognizable immediately by their high, convex carapace that has evolved to contain the enlarged posterior lobe of the venom gland, which occupies much of the prosoma.

Most spiders have eight eyes in two rows, although some families have fewer, or even no eyes at all. The basic arrangement consists of two rows of four eyes each. These eye rows may be straight, recurved or procurved. The eyes are named according to their relative positions on the carapace as follows: anterior lateral eyes (ALE = each outer eye of the front row), anterior median eyes (AME = middle two eyes of the front row), posterior lateral eyes (PLE = each outer eye of the hind row), and the posterior median eyes (PME = middle two eyes of the hind row). The median ocular quadrangle (MOQ) is the area of the four median eyes. The distance between the base of the AME and the anterior edge of the carapace is called the clypeus. The number of eyes is not always distinctive because it may vary within a single family, although this is unusual. Nonetheless, the eye arrangement of spiders (including relative sizes) is often highly diagnostic.

Four pairs of walking legs are attached to the prosoma and are numbered 1–4, from anterior to posterior. Each leg consists of seven segments: the basal segment is called the coxa, which is followed by the trochanter, femur, patella, tibia, metatarsus and the distal-most segment, the tarsus. The tarsus may be pseudosegmented in some families and ends in either two or three claws, with or without an onychium, depending on the family. All web-building spiders have three claws (the third or unpaired claw is used to grasp the silk strand), but not all three-clawed spiders build webs. The leg arrangement relative to the body (prograde versus laterigrade) and the relative leg lengths are often useful diagnostic features. This is also true of the highly variable leg armature, which may consist of several specialized types of setae (hairs) or macrosetae (spines). The former include long, extremely fine sensory hairs called trichobothria, which originate from cuticular sockets. They are very useful for identifying particular spider families or even genera, but can be very difficult to observe even using a microscope. The socket is usually seen before the hair. In material examined under alcohol with a microscope, the trichobothria are usually seen moving with the fluid. Other important features of the legs include clasping spurs (particularly in males of some families), a comb of serrated bristles on the fourth metatarsus in comb-footed spiders (family Theridiidae), the presence or absence of dense scopulae (very fine pads of setae which facilitate adhesion and thus climbing in non-ground dwelling species), claw tufts, and the calamistrum of cribellate spiders.

The short, leg-like appendages anterior to the first pair of legs are called pedipalps. These lack a metatarsus, so have only six segments, with the tarsus of females bearing a single, terminal claw in some families. In the mature male, the tarsus of the pedipalp is specialized to serve as a sperm transfer organ during copulation. These modifications can be either very simple, consisting of a single sclerotized bulb with the tip extruded to form an embolus (haplogyne spiders) or much more complex with the tarsus modified to form a spoon-shaped structure called the cymbium, which holds the genital bulb consisting of several sclerites, the largest of which is usually the tegulum (entelegyne spiders). The tegulum gives rise to the embolus, which may be a long coiled tube in some species, whereas in others it may be short and straight. Other sclerites include the conductor, which supports the embolus, and the median and terminal apophyses. Often the pedipalpal tibia, femur or patella has one or more sclerotized, cuticular projections or apophyses. All of these pedipalpal characters are highly species specific and are usually the most important structures for spider taxonomy. Thus, spider species identification requires adult specimens. The pedipalp of immature males resembles that of females, but in those nearing their final moult (i.e., penultimate instar males) the pedipalps may become swollen but lack the sclerites.

The coxae of the pedipalps are expanded to form the endites or maxillae on either side of the labium, all of which surround the opening to the mouth. Anterior to these structures is a pair of chelicerae, each of which terminates in a hollow fang. These are used to grasp and masticate (not in all families) prey and are sometimes highly modified for use in courtship, mating, or for digging. The fang rests in a furrow when not extended, the margins of which often contain cuticular projections known as cheliceral teeth. The cheliceral dentition of a spider is sometimes useful for identification purposes. Behind the mouthparts and surrounded by the leg coxae is the sternum. A labiosternal groove usually demarcates clearly this sclerite from the labium, although in some families (e.g., Filistatidae) the two are fused. In some families (e.g., some Uloboridae: *Miagrammopes*) the sternum may be divided into separate regions by weakly sclerotized, transverse bands. Precoxal sclerites are sometimes present between the sternum and coxae, and intercoxal sclerites between the coxae.

The opisthosoma consists of the abdomen and associated appendages. The exoskeleton is considerably thinner than that of the prosoma, allowing for expansion during feeding and egg production. In the majority of spiders the abdomen is elliptical, oval or globose, without ornament and hence of little value for the identification process. However, in some species the abdomen may be adorned with spines (e.g., some Araneidae: *Gasteracantha*). In some families (e.g., dwarf hunting spiders, family Oonopidae) the abdomen may contain sclerites known as scuta, which may be positioned dorsally and ventrally. These should not be confused with sigillae, which form internal points for muscle attachment, and are usually visible dorsally as four small, circular depressions. The heart is often visible through the integument dorsally as a longitudinal cardiac mark. The abdomen may also be decorated with patterns consisting of spots, bands, chevrons or a folium, which in some species may include highly conspicuous aposematic (warning) colours, which suggests that the spider may have a venomous bite.

When the abdomen is viewed ventrally, a number of structures are readily visible, including the spinnerets and the openings for the respiratory and reproductive systems. The spinnerets are the silk-producing organs and are usually located near the posterior end of the abdomen (there are a few exceptions). All spiders possess spinnerets but not all spiders weave webs, though they still use silk for other purposes such as constructing retreats, egg sacs, drag lines, dispersal by ballooning, etc. Most spiders have three pairs of spinnerets: the anterior laterals, posterior medians and posterior laterals. The position, number and relative lengths and thickness of the spinneret segments are very useful characters for identifying certain spider families (e.g., Hahniidae, Gnaphosidae, Prodidomidae, Oecobiidae and Hersiliidae). Immediately posterior to the spinnerets is the anal tubercle, which covers the excretory opening. Cribellate spiders have an additional structure located immediately anterior to the spinnerets. This is the cribellum or spinning plate and may be either entire or divided. It produces a special type of silk, which is combed out using the calamistrum on the metatarsus of the fourth pair of legs. This silk appears thick and fuzzy (often with a bluish tinge) and is mechanically, rather than chemically sticky. In the ecribellate spiders a tiny, non-functional vestige called the colulus is often present instead and is considered to be homologous to the cribellum.

Most spiders use two types of respiratory systems: a pair of booklungs and tubular tracheae. The external openings to the booklungs are located in the anterior portion of the abdomen on either side of the epigastric region. The external opening to the tracheae is called the spiracle and is visible as a transverse slit usually located immediately anterior to the colulus or cribellum. Some tiny spiders rely solely on tracheae for respiration and so lack booklungs. The epigastric furrow contains the gonopore (the reproductive opening) and runs transversely across the abdomen between the booklungs. In males, it leads to the testes and is barely visible. In females it leads to the ovaries and in mature individuals is usually associated with sclerotized external structures and is referred to as the epigyne. Immediately beneath the epigyne is the vulva which usually consists of spermathecae and copulatory and fertilization ducts. The structure of the epigyne and the vulva is usually species specific and forms the basis of female spider taxonomy. However, these structures appear to be less useful than the male pedipalp morphology, which forms the basis of most spider taxonomy. A generalized schematic of the internal organs of a spider is shown below.

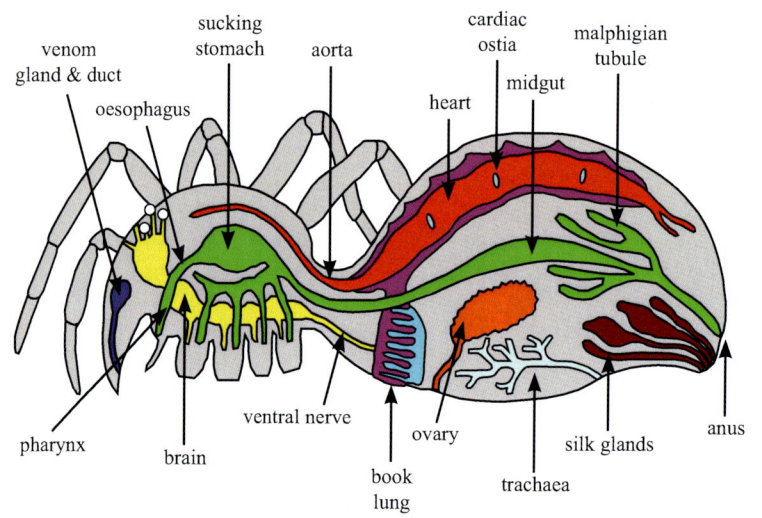

Enemies of Spiders

Species from many different insect orders are specialist predators of spiders. Particularly notable in this regard are the parasitic wasps (Hymenoptera). In South Africa one species is bold enough to snatch the giant *Nephila* spiders from their webs, but this has not yet been observed in The Gambia, although you regularly see webs where the female seems to have been forceably removed! Some parasitic insects paralyse the spider before laying an egg on it and sealing it in a burrow or specially constructed chamber. Others lay their eggs on living spiders and the larva develops attached to the abdomen. Some fly (Diptera) and mantispid (Neuroptera) larvae feed on spider eggs. Certain spiders also prey on other spiders.

Spider hawk wasp with paralysed prey

Opened nest of mud dauber wasp

Fly laying eggs in spider egg sac

Some mantis fly larvae eat spider eggs

Araneidae
(Typical Orb-Web Spiders)

A large and diverse family, most of which weave the vertical orb-web which typifies spiders to scientists and laypersons alike. However, there are modifications. The subfamily Cyrtophorinae construct a permanent, tent-like structure supporting an extremely fine-meshed horizontal web. There is usually a stack of egg sacs, beneath which the spider hangs. These hatch from the top down and the offspring make tiny miniatures of the parental web inside that of the mother, before dispersing after several instars. Most members of this family belong to the subfamily Araneinae, which do not have any major distinguishing features. They generally have a globose abdomen (often with a pattern) that overhangs the carapace and their orb-web usually has no additional decorations. One exception is *Cyclosa*, which has the abdomen longer than wide, projecting beyond the spinnerets, and which also constructs a linear detritus stabilimentum in its web, which consists of egg sacs, shed skins and prey remains. The spider has been observed using this for defensive (camouflage) purposes. Juveniles of *Pararaneus* build a horizontal orb-web, with the centre raised into a conical form by three silk threads attached to the hub. Females of the subfamily Argiopinae are relatively large, have yellow-striped legs, a silver carapace and distinct posterior abdominal humps. Adults usually have conspicuous thick bands of white silk (stabilimenta) radiating from the hub of the web, but in juveniles this is replaced by a circular zig-zag structure. Males of *Argiope* are much smaller and look nothing like the female. Species in the subfamily Gasteracanthinae are easily recognized by their spiny abdomens. They too construct stabilimenta, which consist of small tufts in *Gasteracantha* and fine lines in *Isoxya*. The family includes both diurnal and nocturnal species, most of which build a new web every day (after first consuming the old one). Some of the nocturnal species are highly cryptic during the day and difficult to spot.

Tent-web spider *Cyrtophora citricola*

Tent-web of *Cyrtophora citricola*

Forest orb-web spider *Pararaneus* sp.

Green orb-web spider *Araneus apricus*

Grass orb-web spider *Neoscona* sp.

Cryptic orb-web spider ?*Neoscona* sp.

Orb-web spider *Neoscona* sp.

Male orb-web spider ?*Araneus* sp.

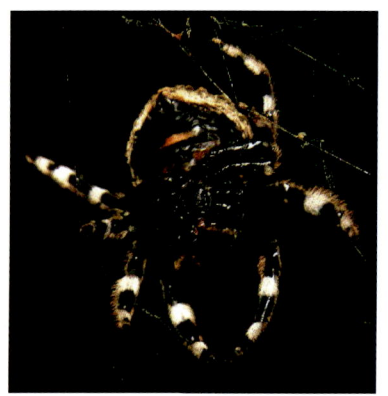

Bark orb-web spider *Caerostris* sp.

Two spot orb-web spider ?*Araneus* sp.

Unusual araneid spider found on leaves

Garbage line spider *Cyclosa* sp.

24

Female zig-zagger *Argiope flavipalpis*

Stabilimentum of juvenile *Argiope*

Male zig-zagger *Argiope flavipalpis*

Spiny orb *Gasteracantha curvispina*

Kite spider *Isoxya semiflava*

Linear stabilimentum of *I. semiflava*

CITHAERONIDAE
(CURLY-LEGGED HOUSE SPIDERS)

Relatively small spiders with a body length of less than 10 mm, they are easily identified by their eye arrangement (posterior median eyes largest and of irregular shape), spineless legs and distinctive spinnerets, which protrude, but are not cylindrical as in the closely related Gnaphosidae. In addition, the last segment of each leg (the tarsus) curls up in dead specimens preserved in alcohol. This is a small family and only the genus *Cithaeron* is known from Africa. Very little information is available on the biology and natural history of these spiders, but they are often referred to as extremely fast, nocturnal ground-running spiders of forest and savannah habitats, that spend the day in silken retreats under stones.

In The Gambia these are one of the most common synanthropic (living with humans) spiders and are often found in houses, either running across the floor, but more often than not high up on a wall. Within Africa, the genus has also been recorded from Ivory Coast, Libya, Egypt, Ethiopia and Yemen. The Gambian specimens belong to the most widespread species in the family *Cithaeron praedonius*, which has also been reported from Europe (Greece), Malaysia, Australia and most recently from Brazil, where they were also found to have synanthropic tendencies.

The synanthropic *Cithaeron praedonius*

CORINNIDAE
(SAC SPIDERS)

This is a diverse, but poorly delimited family. Most are less than 10 mm in body length and they range in colour from dull brown to bright orange. Corinnids have eight eyes in two rows of four, which may be either widely spaced or closely grouped. They do not weave webs and are usually found in leaf litter, particularly in forests. Many species mimic ants or other insects such as mutilid wasps (velvet ants). This is probably a rather diverse family in The Gambia, but little more can be said in this regard without a comprehensive taxonomic and systematic study. Without careful examination, in the field they may be mistaken for small Ctenidae or Lycosidae.

Common corinnid *Messapus martini*

Striped corinnid *Copa* sp.

Ant mimic corinnid *?Apochinomma* sp.

White-tail *Castianeira loricifera*

CTENIDAE
(WANDERING SPIDERS)

Small to very large nocturnal wandering spiders that hunt prey at ground level and on vegetation, resting under logs and within leaf litter during the day. They are easily identified from the similar Lycosidae and Pisauridae, by their eye arrangement of eight in three rows, with the anterior laterals tiny and situated between the posterior medians and posterior laterals. The tibiae of the anterior legs usually have numerous pairs of ventral spines. In some parts of the world wandering spiders are extremely aggressive, bite readily and are poisonous to humans. This would appear not to be the case in Africa, but Gambian specimens should be handled with care just in case.

?new species, common in leaf litter

Spotted wandering spider *Anahita* sp.

Striped wandering spider ?*Ctenus* sp.

Wandering spider ?*Africactenus* sp.

Deinopidae
(Net-Casting Spiders)

Deinopids are easily identified by their huge posterior median eyes (giving them their other common name of ogre-faced spiders) and unique predatory behaviour. They hunt in the evening suspended from a framework of threads, where they construct a highly elastic capture web, which they hold in their front legs. When a prey item crawls beneath (or flies past) the snare is lunged outwards and expanded. If contact with the prey is made the spider jerks the web several times in order to make sure it is firmly trapped. They are highly cryptic during the day and are extremely difficult to find, although sometimes you may encounter an unused capure web suspended in vegetation. The spider will certainly be close by waiting for nightfall.

Net casting spider *Deinopis* sp.

DICTYNIDAE
(LEAF MESH-WEB SPIDERS)

This is the largest family of spiders which possess a calamistrum and cribellum, with more than 500 species worldwide, but they are more common in temperate regions than they are in the tropics. The limits of the family as currently defined are erroneous and much work is required to resolve this issue. However, in general, they are small spiders, usually less than 5 mm in body length, which are usually pale in colour, but with a dark pattern on the abdomen. Species belonging to the genus *Nigma* are a uniform, pale green, whereas others have a reddish-brown ground colour.

Little is known about the behaviour or ecology of West African Dictynidae, or indeed how many species occur in the region. They build messy, irregular mesh-webs at the tips of plants or across leaves, where they also lay their egg sacs, and sometimes females may be found co-existing with their offspring. Whether there is any additional degree of maternal care is unknown (note the two spiderlings in the top right corner of the bottom left photograph).

In addition to *Dictyna* and *Nigma* (shown in the photographs below and here recorded from The Gambia for the first time), *Mizaga chevreuxi* has also previously been recorded from the country as well as from Senegal.

Black and white dictynid *Dictyna* sp.

Green dictynid *Nigma gertschi*

Eresidae
(Velvet Spiders)

These spiders derive their name from the dense layer of short setae that cover the body and legs, giving them a velvety appearance. They rarely exceed 15 mm in length and have eight eyes, with the laterals well separated. The legs are relatively short and stout, with the tarsus and metatarsus forming a single functional unit. *Stegodyphus* is usually found on plants and two of the sixteen known African species form colonies, but this phenomenon has not been observed in The Gambia. A second Gambian species (the white spider in the photograph) is common by the coast and constructs a messy web, often within an abandoned crab burrow or beneath fallen palm fronds.

Velvet spider *Stegodyphus* sp. 1

Velvet spider *Stegodyphus* sp. 2

Stegodyphus sp. 2 web in fallen frond

Stegodyphus sp. 2 spiderlings

FILISTATIDAE
(CREVICE-WEAVING SPIDERS)

These small (10 mm) nocturnal spiders are common in the more arid, hotter regions of the world. They live in silk-lined retreats, with messy radiating webs of cribellate signal lines. They are usually located within an existing crevice, and are also very common on the outside walls of compounds. Sometimes they are found wandering around inside houses, particularly males as they search for a mate.

The carapace is pointed anteriorly and there are eight eyes in a small group, situated on a slightly raised mound. This family is unusual in that the labium is fused to the sternum, and the spinnerets are situated slightly forward. Both males and females hold their pedipalps at an odd, raised angle when at rest. Although these are cribellate spiders, possessing both a cribellum and a calamistrum, the latter can often be very difficult to see, because it is rather short and consists only of rows of disordered setae.

This family is thought to be one of the most primitive of the haplogyne spiders (those with simple reproductive organs) and various regional studies have been conducted on these spiders for Australia, Asia and South America. At least three different genera have been recorded from West Africa, but the Afrotropical region is still awaiting revisionary studies.

Crevice weaver *Afrofilistata fradei*   Web of *Afrofilistata fradei*

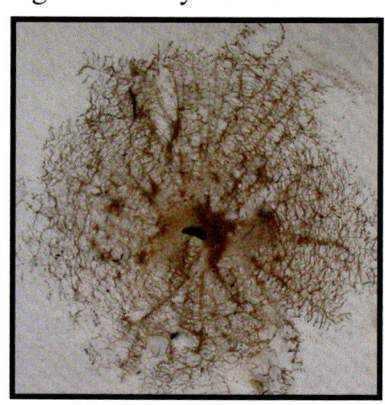

Gnaphosidae
(Flat-Bellied Ground Spiders)

One of the hyperdiverse families with approximately 2,000 described species worldwide. These small to medium-sized spiders are easily recognized by having eight eyes in two rows, usually with irregularly shaped posterior median eyes and well separated, large, parallel, cylindrical anterior spinnerets (but without greatly elongated spigots as in Prodidomidae). The legs are usually spinose, short and stout, in contrast to Cithaeronidae, which have similar features with regard to eye structure and to a lesser extent spinneret morphology. The fourth leg is usually the longest and there may be a preening comb (a group of specialised spines) on the metatarsus.

Gnaphosids are generally considered to be free-living spiders that live at ground level, although they can also be found well off the ground. They come out to hunt at night after spending the day in silken retreats beneath stones or other objects. More than 300 species have been described from the Afrotropical region. Most are rather dull in colour, but the shiny black colour of *Zelotes* is quite striking. They tend to be more common in dry, open habitats rather than in forested areas, although they do occur there. They run quickly when disturbed and are difficult to catch. Some South African species prey specifically on termites, others are common in houses.

Shiny black ground spider *Zelotes* sp.

Ground spider *Minosia* sp.

HAHNIIDAE
(COMB-TAILED SPIDERS)

These tiny spiders are usually less than 5 mm long and are easily recognized by having their spinnerets positioned in a single, wide, transverse row, with the posterior pair long and two-segmented (they look like the teeth of a comb, giving the spiders their common name). They are usually found in forested areas, where they construct delicate sheet-webs close to the soil surface, sometimes in animal footprints. They lie in wait beneath grains of soil close to the webs and can also be found in moss on tree trunks. These spiders are most easily found early in the morning when the dew-coated webs become clearly visible.

The family has a worldwide distribution with only two genera recorded from Africa. The Gambian species probably belongs in the genus *Hahnia*, which is known to occur throughout the Afrotropical region. The other genus recorded from the continent is *Muizenbergia*, which includes one intertidal species, found on the rocky coasts of South Africa. They live inside the calcareous masses constructed by marine annelid worms and their bodies are covered with water-repellent hairs. These form a plastron (a physical gill), which traps air and allows the spiders to breathe when submerged under water.

Comb-tailed spider *Hahnia* sp.

HERSILIIDAE
(TWO-TAILED SPIDERS)

Easily recognized by their extremely long spinnerets which give them their common name. They usually live on tree trunks and can be very difficult to spot (males more so than females), not least because they often run around to the other side as you approach, returning to their original resting position a short time later. Females sit above their egg sacs which they disguise with bits of debris. I am of the opinion that these spiders can change colour to match the substrate on which they are resting (as is known to happen in some other spider families), but that this takes some time to achieve. However, this idea has yet to be tested. There is probably only one species in The Gambia.

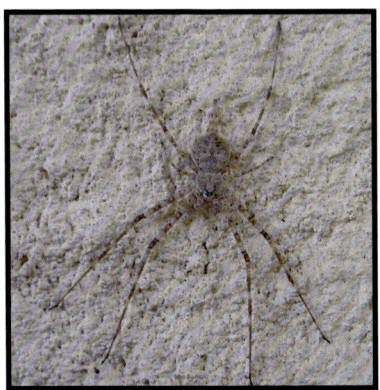

Two-tailed spider *Hersilia caudata*

Highly cryptic *H. caudata* on tree trunk

H. caudata guarding egg sacs

Dark form of *H. caudata*

Linyphiidae
(Dwarf Spiders)

This is one of the largest spider families, but they tend to be more common and diverse in northern, temperate countries, such as those in northern Europe. They resemble comb-footed spiders (family Theridiidae), but lack the comb of serrated setae on the tarsus of the fourth leg. Both morphology and behaviour are variable and most species are tiny and so are easily overlooked. The pedipalps of mature males are usually large and complicated, and males often have stridulatory ridges on the outside surface of the chelicerae.

There are two main subfamilies: Linyphiinae (hammock-web spiders) and Erigoninae (money spiders). The former usually have longer legs and hang upside-down in domed or hammock-shaped webs, which are usually supported by silk strands above and below. The latter are mostly free-living in leaf litter or under surface debris, but some also construct tiny, flat sheet-webs close to the ground.

At least 76 genera have been recorded from the continent, with almost half of these occurring in West Africa. How many of these are present in The Gambia is unclear because there are no Gambia specific records. It can also be expected that The Gambia will provide new records of species and genera previously unrecorded elsewhere in Africa.

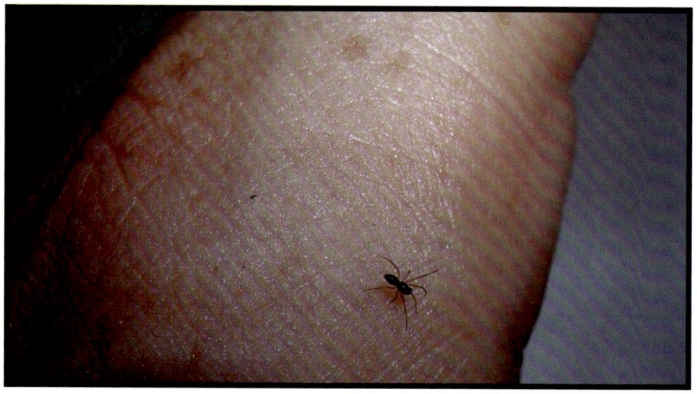

An adult linyphiid on the author's finger—they really can be very small!

LYCOSIDAE
(WOLF SPIDERS)

This is an extremely diverse family with more than 50 genera in at least six subfamilies recorded from Africa. They range in size from small to large and are easily identified by their highly characteristic eye arrangement of eight eyes in three rows, with a pair of large posterior median eyes set above a row of four small anterior eyes. Behaviour also facilitates identification: females carry their egg sacs fixed to the spinnerets and the young ride on their mothers' back for a period after hatching.

Most are free-living, ground dwellers with both diurnal and nocturnal species, the latter often spending the day hiding in deep, silk-lined burrows or under stones. They tend to capture their prey using a sit-and-wait ambush strategy, rather than chasing them, as their name might suggest. These spiders are abundant in open savannah and agroecosystems, where they play an important role in the control of insect pests. Some diurnal species e.g., *Pirata* are found close to water and can even skate across the surface film. Larger species such as the nocturnal *Geolycosa* and *Lycosa* are burrowing spiders. *Hippasa* (common in Abuko) are unusual for wolf spiders in that they construct sheet-webs with funnel retreats. At least nine different wolf spider genera have been recorded from Senegal, but there are no previous Gambian records.

Typical habitus and eye arrangement of wolf spiders

Wolf spider *Hippasa cinerea*

Funnel-web of *Hippasa cinerea*

Burrowing wolf spider *Geolycosa* sp.

Burrow of *Geolycosa* sp.

Large wolf spider *Lycosa* sp.

Wolf spider carrying egg sac

Miturgidae
(Prowling Sac Spiders)

This is a rather diverse family, with four subfamilies recorded from the Afrotropical region. However, only the genus *Cheiracanthium* (subfamily Eutichurinae) is known to occur in The Gambia, although several others occur in West Africa, including Senegal. It has been suggested that *Cheiracanthium* is misplaced in Miturgidae. These spiders have a body length of about 10 mm, have the first pair of legs longer than the others and are pale green-brown with dark, shiny chelicerae. They are free-living, nocturnal predators, usually found on vegetation and often inside houses. *Cheiracanthium* lacks a fovea and the posterior spinnerets are noticably longer than the others.

They construct sac-like retreats and the female encloses herself with her eggs in a folded blade of grass to protect them as they develop. They are aggressive and bite readily. People are sometimes bitten whilst sleeping, but also when they dress in clothes where a spider may have made its retreat. The bite is often painless at first, but there will be two puncture marks 4–6 mm apart. The area later becomes inflamed and develops into an ulcerating lesion within a few days, which can be large and take a long time to heal. It has been suggested that a single species of *Cheiracanthium* (*C. furculatum*) is responsible for 90% of all human spider envenomations in South Africa.

Prowler *Cheiracanthium furculatum* *C. furculatum* guarding eggs

MYSMENIDAE
(SPURRED ORB-WEAVING SPIDERS)

These tiny spiders have a body length of less than 3 mm and the carapace usually has a raised ocular area with eight eyes in two rows. The pedipalps of mature males are large and there is a mating spur on the underside of the metatarsus of each first leg (both visible in the photo). Females have a small sclerotised, ventro-subdistal spot on the femur of the first pair of legs (visible on the male in the photo: arrow). The tarsi are equal in length to (or longer than) the metatarsi.

There has been considerable scientific debate regarding the correct systematic placement of these spiders and over the years they have been transferred from one family to another, including Theridiidae, Symphytognathidae and Anapidae. It has also been suggested that the Old World genera (which includes the Gambian species) should be transferred to the Synaphridae, which has only recently been given family status.

Little is known about the biology and behaviour of mysmenids, although some are known to be kleptoparasites in the webs of other spiders. However, the species illustrated, which is quite possibly new to science (the genus having previously been recorded in Africa only from Rwanda) probably constructs a small orb-web in the humid leaf litter layer of forests.

Spurred orb-weaver *Mysmenella* sp.

NEPHILIDAE
(GIANT ORB-WEB SPIDERS)

These huge golden-silk orb-weaving spiders are unmissable during the rainy season and quite unmistakable. The females are huge (body length 40 mm), but the males are tiny. *Nephila fenestrata* prefers shaded habitats such as forests, whereas *N. senegalensis* prefers open areas and is common in urban settings, where they may form large 'colonies'. Both species construct huge orb-webs, which are usually made of golden silk, particularly in more mature individuals. The purpose of the colour in the silk is unknown. The webs often contain several males, with the dominant individual on the other side, opposite the female. He often waits until she is feeding before attempting to mate, in order to reduce the risk of becoming a meal himself. Small kleptoparasitic spiders (e.g., Theridiidae: *Argyrodes* sp.) are often found, sometimes in large numbers (25+). *N. senegalensis* lays her egg sacs above ground, either in vegetation or attached to walls, whereas those of *N. fenestrata* are buried. Both hatch during March/April with adults conspicuous throughout the rainy season, but most of them have died off by the mid-dry season.

Nephilids have only recently received family status and will be found under either Araneidae or Tetragnathidae in older spider related publications.

Forest golden-orb *Nephila fenestrata*   Spotted golden-orb *N. senegalensis*

N. fenestrata mating

Sexual cannibalism in *N. fenestrata*

Cluster of *N. fenestrata* spiderlings

Juvenile *N. fenestrata* with prey

Large 'colony' of *N. senegalensis*

Egg sacs of *N. senegalensis*

OCHYROCERATIDAE
(TINY GROUND-WEAVING SPIDERS)

These long-legged spiders are minute, with a body length of approximately 1 mm. They have only six eyes, with an anterior row of four and a posterior row of two widely separated posterior laterals. The chelicerae have a median lamella and the legs have three tarsal claws situated on an onychium.

Little is known about the biology and behaviour of these inconspicuous, cryptozoic spiders, which are found in tropical regions worldwide. Some species live in caves, but in The Gambia they are found in the humid leaf litter layer of forests, where they presumably construct irregular 3-dimensional space-webs. They are also common between the leaf bases and trunks of rhun palms, where a similar litter-type micro-habitat develops over time. Females produce small clusters of eggs which are loosely held together with silk and carried in their chelicerae, appearing as a small ball in front of the spider.

Some species from this family are known to be parthenogenetic, i.e., the female is able to produce viable eggs and offspring in the absence of a male. Whether or not this is the case for the Gambian species is unknown. However, given that females (with eggs) can occur in very large numbers and not a single male has been found to date in The Gambia, this is a possibility that cannot be ruled out.

?*Speocera* sp.

OECOBIIDAE
(DWARF STAR-LEGGED SPIDERS)

These tiny cribellate spiders have a maximum body length of only 3 mm. They are easily identified by their characteristic star-legged appearance, and microscopically by their subcircular carapace with eight closely grouped eyes. The carapace has a distinct clypeal snout, and they have elongated posterior spinnerets and a large anal tubercle, which is fringed with long, curved setae. The colour, in terms of abdominal pigmentation and leg banding can be rather variable.

They are usually found in reasonably large numbers on the outside walls of buildings, where they rest beneath small, star-shaped mesh-webs (remarkably similar in structure to the millenium dome in London!). Males are often found together with females. The prey, usually an ant, is immobilized by the spider rapidly circling it and enswathing it with silk. These spiders are extremely fast runners when disturbed and so can be difficult to catch.

The African oecobiid fauna consists of both indigenous species of limited distribution and some synanthropes, which also occur on other continents. The Gambian species (also known to occur in Senegal, Mali and Nigeria) falls into the former category, although it is also synanthropic, being found in and around human dwellings.

Star-legged spider *Oecobius pasteuri*   Female (left) and male *O. pasteuri*

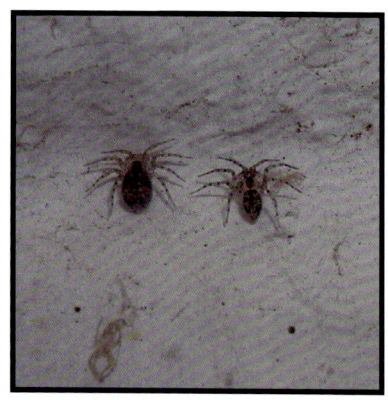

Oonopidae
(Dwarf Hunting Spiders)

This diverse family of very small (less than 3 mm) spiders are found throughout the tropics and to a lesser degree in temperate regions. They are nocturnal, free-living hunters that are extremely common in forests on the soil surface below the leaf litter layer. They may also be found on tree trunks. In contrast to most other spiders, they have only six eyes, which are closely grouped. Two subfamilies have been recorded from Africa and both occur in The Gambia. Gamasomorphinae are easily identified by having dorsal and ventral abdominal scuta (sclerotized plates), which are absent in Oonopinae. Most species appear shiny and are often orange-red in colour.

Several species have been found on the soil surface beneath moist leaf litter in Gambian forests (Bijilo, Abuko, Tanji Bird Reserve), including the genera *Ischnothyreus* (Gamasomorphinae) and *Orchestina* (Oonopinae), with others yet to be identified. *Orchestina* are easily identified by having the femora of the last pair of legs distinctly swollen. This appears to be an adaptation for jumping. As yet, no species have been recorded from Senegal, but will certainly occur there. In other African countries (e.g., Zaire) some species with no eyes are specialists that live in termite nests. A worldwide revision of the family is currently underway.

An unidentified oonopid spider

Oxyopidae
(Lynx Spiders)

This is a relatively large spider family in terms of numbers of species, which seem to be rather common and diverse in Africa and also within The Gambia. Unfortunately, very little taxonomic work has focussed on the African fauna other than a revision of the genus *Peucetia*. Most genera have a body length of less than 10 mm, but the bright green *Peucetia* species may be twice that size. Members of this family are easily identified by their extremely long leg spines and by having eight eyes closely grouped in a hexagonal arrangement with a high clypeus.

They are free-living hunters that are usually found on vegetation, but are also common on walls of buildings. Many species are diurnal, have good vision and have been observed leaping off vegetation to capture insects in flight. They are fast, agile and jump readily, which may cause them to be mistaken for jumping spiders (family Salticidae) at first glance. Females construct their egg sacs adpressed to leaves or suspended in a small irregular web. The females remain with the eggs until they hatch, and at least in *Peucetia*, they remain with their offspring for a short time. Several species rest with their first three legs held close together and directed forwards and can be highly cryptic and difficult to see. When alarmed, *Hamataliwa* raise their first pair of legs in what appears to be an aggressive manner, presumably aimed at warding off a would be predator. Other species leap away rapidly.

At least three different genera occur within The Gambia: *Peucetia*, *Oxyopes* and *Hamataliwa*, but exactly which species they are will remain unknown until revisionary studies that include Gambian specimens are conducted. The Afrotropical *Peucetia* were revised in 1994, but no Gambian material was available for comparative research purposes. *Peucetia striata* has the most widespread distribution on the continent, but this new Gambian record represents the first one for West Africa.

Green lynx *Peucetia striata*

Tiger lynx *Hamataliwa* sp.

Brown lynx *Oxyopes* sp.

Orange lynx *Oxyopes* sp.

Lynx with egg sac adpressed on leaf

Lynx with suspended egg sac

47

Palpimanidae
(Palp-Footed Spiders)

These highly distinctive spiders are easily recognized by their red colour and enlarged first pair of legs with thick scopulae (dense hair pads) on the inner margins of the tibiae, metatarsi and tarsi. They are usually less than 10 mm in body length, but are secretive and seldom seen, although they can be relatively abundant in forests beneath the leaf litter layer. Palp-footed spiders do not weave capture webs and very little is known about their behaviour, although it is suspected that many species prey on other spiders. When at rest, they sit with their powerful front legs held close to the body. When foraging for food they hold the first pair of legs forwards and angled slightly outwards with the tips of the tarsi close to the floor. This action presumably funnels in potential prey items as the spider moves forwards. When alarmed, these spiders raise their powerful front legs upwards in a threatening manner.

This is the first record of the family from The Gambia, although the monotypic genus *Badia*, known only from the female, has been recorded from Niokolo Koba, Senegal. Four species of *Sarascelis* occur throughout West Africa, with records from Ivory Coast, Guinea-Bissau and Ghana. My suspicion is that the Gambian species will turn out to be *Sarascelis chaperi* as I was unable to identify it as any of the other species.

Palp-footed spider *Sarascelis chaperi*

PHILODROMIDAE
(RUNNING CRAB SPIDERS)

These agile, free-living hunting spiders look like smaller versions of the giant crab spiders (family Sparassidae), which are common in houses, although they seldom exceed 15 mm in body length. They also resemble some members of the true crab spiders (family Thomisidae) because of their laterigrade legs and have been considered as a subfamily therein. However, philodromids have scopulae on the tarsi of the first two pairs of legs, whereas thomisids do not. In addition, the eyes are never on large tubercles as they are in Thomisidae. Running crab spiders are usually found on the soil surface or on vegetation.

The philodromids of the Afrotropical region are not well known, but at least six genera have been recorded from West Africa, but these represent the first records for The Gambia. *Thanatus* is usually found on the soil surface, whereas *Philodromus* is common on vegetation or on tree trunks. *Tibellus seriepunctatus* has been recorded from Senegal and this genus is common throughout savannah regions, but they are cryptic spiders with elongate straw-coloured bodies and rest on upright grass stems with their legs held close together, making them very difficult to spot. I have only observed very small specimens, but sampling with a sweep net or beating tray would certainly yield adults to confirm the identification.

Running crab spider ?*Philodromus* sp.   Running crab spider ?*Thanatus* sp.

Pholcidae
(Daddy-Long-Legs Spiders)

These small to medium-sized spiders are easily identified by their extremely long legs with flexible tarsi and no tarsal comb on the hind legs. Their eye arrangement consists of two triads and usually a small pair of anterior medians. They weave irregular space-webs and are particularly common in buildings (*Crossopriza* and *Micropholcus*) and in tree holes and beneath logs in forests (*Smeringopus*). *Leptopholcus* is usually found resting flat against the underside of leaves. They vibrate rapidly in the web when disturbed and females carry their eggs in their chelicerae. In addition to those species shown, *Pehrforsskalia conopyga* has been observed in Abuko Nature Reserve.

House phoclid *Micropholcus fauroti*

Forest pholcid *Smeringopus pallidus*

Large house phoclid *Crossopriza lyoni*

Leaf pholcid *Leptopholcus guineensis*

Pisauridae
(Nursery-Web Spiders)

This is a large and diverse family within the Afrotropical region, with at least 32 genera known to occur there. Four species have previously been recorded from Senegal, but none from The Gambia. The family is ill-defined, in that it has no known unique morphological features, and to date, the nursery-web, a web built by the female for the freshly hatched spiderlings has been their only defining feature. Hence the common name for the family of nursery-web spiders. Females often remain by their nurseries and guard the young until they disperse. However, behavioural data for many of the genera are scarce and less than one-fifth of the recognized genera are known to weave such a web.

Most are medium-sized, but some can reach up to 30 mm in body length. They have eight eyes in two, three or four rows, an elongate abdomen that tapers towards the rear and often has a median longitudinal band (also present on the carapace), and long, slender legs with three tarsal claws. The females construct spherical egg sacs, which they hold in their chelicerae and pedipalps and carry around beneath their sternum.

Little is known about the biology of Afrotropical nursery-web spiders, but their life-styles are varied, with some weaving fine sheet-webs and others free living on vegetation or on the ground. Some species are associated with freshwater habitats and are even able to catch fish and amphibians. They can also move across the surface film, swim and dive underwater to escape from predators. Males of one European species are known to offer silk-wrapped prey items as prenuptial gifts to females before attempting to mate, but whether this behaviour occurs in any of the African species is unknown.

Many of the Afrotropical genera have been revised, albeit mainly in the 1970s, but the identification of some genera is still problematic due to the inadequate treatment of this family in an earlier work.

Unidentified female with egg sac

Unidentified female with nursery web

Free living pisaurid ?*Perenethis* sp.

Free living pisaurid ?*Perenethis* sp.

Web pisaurid ?*Rothus* sp.

Web pisaurid *Euprosthenopsis* sp.

PRODIDOMIDAE
(PRODIDOMID GROUND SPIDERS)

These tiny to medium-sized (up to 10 mm) spiders are easily recognized under magnification by having eight eyes in a characteristic subcircular arrangement and by having well developed anterior lateral spinnerets with elongate piriform gland spiggots.

The specimen in the left photograph was found during the night, high up on an inside wall during the rainy season. It was extremely fast when disturbed and for a so-called ground spider demonstrated remarkable agility on a vertical surface and even upside-down on the ceiling. Most African prodidomids are thought to prefer open, arid habitats, where they are usually found under stones during the day, although *Zimiris* (a tiny spider also recorded from West Africa: Ivory Coast) is known to have synanthropic tendencies (it lives in houses) and would not be unexpected within The Gambia.

Prodidomus is widespread in the Afrotropical region, with previous West African records from Senegal and Guinea-Bissau. Little is known about the behaviour of these spiders, although the taxonomy of the African representatives is relatively well resolved. It has been suggested that some species are associated with social insects because they have been found within the nests of ants and termites.

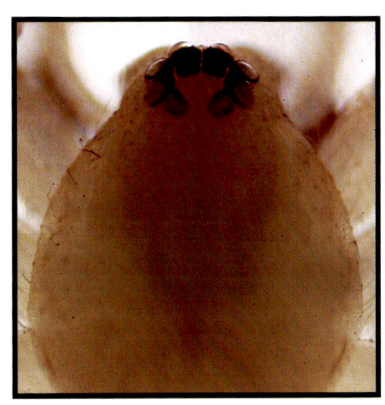

Red prodidomid *Prodidomus purpureus*   Eye arrangement of *Zimiris*

SALTICIDAE
(JUMPING SPIDERS)

Jumping spiders are the most biodiverse spider family on the planet, with more than 5,200 described species in 563 genera. In excess of 600 species are known from Africa. They are often brightly coloured and are easily recognized by their large front eyes, which have excellent colour vision. Many jumping spiders have evolved highly elaborate courtship behaviours, probably due to their excellent visual acuity. Salticids do not weave capture webs (although some hide in silk retreats) and hunt their prey during the day using a stalk and pounce strategy. Most species eat insects, but some feed on other spiders e.g., *Portia* and *Holcolaetis*, with the former known to reason as to the best way to approach their prey. Others mimic ants e.g., *Myrmarachne* and can be difficult to separate from the model.

Most salticids are more or less generalist predators of insects, although there are some pronounced examples of jumping spiders that have specialized preferences and exhibit prey-specific capture behaviours. In Kenya, *Evarcha culicivora* is able to detect, and feeds preferentially on human blood-fed mosquitoes! In The Gambia, *Menemerus bivittatus* is a versatile predator with a repertoire of strategies, which differ depending on the type of prey. For example, stingless bees construct their nests inside cavities on tree trunks, with telltale entrance tubes of wax jutting out from the surface. There is constant movement of individuals entering or leaving the nest and it is not uncommon to find *M. bivittatus* loitering close to the entrance watching the bees as they come and go, but no other salticid species have been observed doing this. This spider is also common on outside walls where it can be observed stalking flies, which are by no means easy to catch. The spider circles around until it is directly behind the fly so that it stands a better chance of approaching the prey undetected. At least 20 species of jumping spiders are known to occur in The Gambia, but there will certainly be many more than this.

Forest jumper *Bianor albobimaculatus*

Red jumper *Cyrba simoni*

Common jumper *Menemerus bivittatus*

M. bivittatus at bee nest entrance

Spider-eating jumper *Portia africana*

Semaphore jumper *Meleon solitaria*

55

Bark jumper *Holcolaetis vellerea*

Holcolaetis vellerea eating a sparassid

Elegant jumper *Evarcha* sp.

Horned jumper *Hyllus* sp.

Masked jumper *Hyllus* sp.

Tiny litter jumper *Stenaelurillus* sp.

Smiling coast jumper ?*Parajotus* sp.

Ant mimic jumper *Myrmarachne* sp.

Striped jumper *Evarcha mustela*

Garden jumper *Thyene* sp.

Flame-back jumper *Natta horizontalis*

White ghost jumper ?*Thyene* sp.

SCYTODIDAE
(SPITTING SPIDERS)

These nocturnal spiders (up to 10 mm) are easily recognized by their high carapace domed in the thoracic region. They have six eyes in three diads, long slender legs and a yellowish body with dark stripes and spots on both the carapace and abdomen. They utilize a remarkable prey-capture strategy. Their domed carapace contains muscles and enlarged poison glands that produce a mixture of venom and sticky silk, which they spit at their prey. They are accurate up to a distance of approximately 2 cm. At least two *Scytodes* species occur in The Gambia. One is common in houses and the other is found on vegetation, where it hides in a curled leaf retreat during the day.

Forest spitter *Scytodes* sp. with eggs

Forest spitter *Scytodes* sp. male

House spitter *Scytodes* sp.

House spitter with phocid prey

SEGESTRIIDAE
(TUBE-WEB SPIDERS)

Segestriids are seldom seen because they are nocturnal and rarely venture out of their retreats. However, they are easily recognized by having elongate bodies (both the carapace and abdomen are longer than they are wide), the third pair of legs directed forwards and six eyes in three groups of two (diads). They are usually dull coloured, medium-sized spiders reaching approximately 10 mm in body length.

These spiders construct silken tubes in any available crevice, usually in bark or on walls. The specimen below used an abandoned termite tube on the trunk of a palm tree, so was easy to exticate and photograph. The silken tube retreat is closed at the inner end, but open at the entrance. A number of dry trip lines radiate outwards from a small collar of white silk at the entrance and serve to alert the spider of passing arthropods. The spider is able to rush out of the tube at a remarkable speed, capture the unfortunate prey and return to safety just as quickly.

No systematic studies of the African fauna have been conducted, but the Gambian species represents a new family record for the country and belongs in the genus *Ariadna*, which is common throughout the Afrotropical region, although it has not yet been recorded from Senegal.

Tube-web spider *Ariadna* sp.   Tube-web spider retreat and webbing

SELENOPIDAE
(FLAT CRAB SPIDERS)

Selenopid crab spiders, also known as flatties are easily recognisable by the incredible dorso-ventral compression of their bodies. They have laterigrade legs and their eyes are arranged in two rows, with the anterior one consisting of six and the posterior of two. They range in size from small to large, with some specimens attaining a leg span of 8 cm. Their extreme flatness means that they can fit into the narrowest of cracks and they are often found in houses, where they spend the day hidden behind wall hangings or pictures, etc. or beneath bark on tree trunks or logs. Adults are nocturnal and rapidly run away sideways when disturbed, although juveniles may be seen during the day.

This family has received recent taxonomic treatment, although data for West Africa are sparse. Only the genus *Selenops* is known to occur in the region, but many species are known from either only the male or only the female, making correct identification difficult in many cases.

Their secretive, nocturnal habits mean that little is known about their behaviour, but they must certainly be important in the control of common household pests, possibly including mosquitoes, in their early instars and cockroaches as adults. Thus, they should be welcome household guests!

Flat crab spider *Selenops* sp.

Flat crab spider *Selenops annulatus*

SPARASSIDAE
(GIANT CRAB SPIDERS)

These large, nocturnal wandering spiders are found on plants, tree trunks and also in houses, where they prey on pests such as cockroaches. *Heteropoda venatoria* is a pantropical species, which is very common in forests but is unlikely to be seen unless you are there at night. Many species e.g., *Olios* spend the day hidden in a silken cocoon. They are very fast when disturbed and some may be mistaken for Selenopidae at first sight, but they are nowhere near as flat and also hold their legs slightly differently. There is a good key to the African genera of this family, but due to little revisionary work the correct identification of species remains problematic.

Giant huntsman *Heteropoda venatoria*

Heteropoda venatoria with egg sac

Female common huntsman *Olios* sp.

Male common huntsman *Olios* sp.

TETRAGNATHIDAE
(HORIZONTAL ORB-WEB SPIDERS)

Also known as four-jawed spiders because of the enlarged chelicerae in *Tetragnatha* species (especially the males), these spiders weave horizontal orb-webs, usually close to water (*Tetragnatha* sp.) or in bushy vegetation (*Leucauge* sp.). The spiders hang upside-down at the hubs. *Tetragnatha* species are dull brown-grey, with elongate abdomens and long legs. They sometimes rest on vegetation with the body and legs held flat against the surface, making them difficult to spot. *Leucauge* species have a silvery abdomen, often with red, green and gold markings; they also have a double row of long trichobothria on the posterior femora (visible under the microscope).

Four-jawed spider *Tetragnatha* sp.

Tetragnatha sp. eating damselfly

Horizontal orb-weaver *Leucauge* sp.

Silver-backed orb-weaver *Leucauge* sp.

THERIDIIDAE
(COMB-FOOTED SPIDERS)

A large and diverse family, which has been poorly studied in West Africa, with no species previously recorded from The Gambia and only five from Senegal. They are small to medium-sized and most weave tangled mesh-webs. They are best identified by the presence of a comb of serated setae on the tarsus of the hind leg (this may be absent in some males).

This is one of the most notorious of spider families because it contains the much feared widow spiders. Brown widows are very common and can be found in every household, whereas the more poisonous black widows are less frequently encountered. However, they rarely bite and there have been no confirmed, recorded fatalities from either species. Nonetheless, they may cause severe and distressing symptoms, especially in children and the elderly.

Argyrodes argyrodes are kleptoparasitic in the webs of other spiders, especially the giant golden orb-weaving Nephilidae. Towards the end of the *Nephila* season I have observed webs harbouring in excess of 25 individuals. They have also been observed consuming the eggs of *Cyrtophora citricola* (Araneidae). Theridiid genera identified from The Gambia to date include: *Latrodectus, Theridion, Argyrodes, Coleosoma, Steatoda, Achaearanea, Euryopis* and *Nesticodes*.

Juvenile black widow *Latrodectus indistinctus* upperside and underside

Brown widow *Latrodectus geometricus* (male and female with eggs/juvenile)

Common comb-foot *Theridion* sp.    House comb-foot *Nesticodes rufipes*

Kleptoparasite *Argyrodes argyrodes* in *Nephila* web and eating *Cyrtophora* eggs

THOMISIDAE
(CRAB SPIDERS)

This extremely diverse family, in terms of numbers of species, morphology and behaviour and are well represented in The Gambia. They are small to medium-sized wandering, sit-and-wait predators that do not weave capture webs. Most are active during the day. Their lateral eyes are usually on tubercles and the first two pairs of legs are often much longer and stronger than the posterior ones. They are usually found on vegetation, but some *Xysticus* live under stones. Many species are highly cryptic on flowers and very difficult to spot.

Many species are colourful and some e.g., *Thomisus* are thought to change their body colour to match their background, although this may take several days and is a reversible process (as observed in South African species). The bright orange and black *Platythomisus* mimics a wasp. They even have two black spots at the front of the carapace to give the impression of compound eyes. When disturbed they hold the black forelegs outwards at right angles to the body and vibrate them up and down to give the appearance of having wings. *Amyciaea* (seen in Abuko Nature Reserve) are convincing mimics of the ants on which they prey. *Tmarus* are usually found on bark and are difficult to see when at rest in their cryptic posture.

Information on the biology of this family is limited. Sexual dimorphism is pronounced in many species and the larger female may even carry the male around on her abdomen, while he waits for her to become receptive. Females usually construct a brood chamber in vegetation for their egg sacs.

They often capture insects much larger than themselves, aided by powerful venom and the strong front legs. Prey are held in a natural position while consumed in order to avoid attracting attention from potential predators that may eat the spider. If you see a butterfly that does not fly away as you approach, look carefully and you will probably find a crab spider attached to it! Thomisids are very useful for controlling insect pests of crops.

Crab spider *Tmarus* sp. eating (note tiny milichiid jackal flies) and in resting pose

Bark crab spider ?*Xysticus* sp.

Ground crab spider *Xysticus* sp.

Wasp mimic spider *Platythomisus* sp.

ant mimic spider *Amyciaea hesperia*

Flower crab spider *Thomisus* sp.

Thomisus sp. feeding on butterfly

Thomisus sp. tiny male on female

Pea spider *Thomisops senegalensis*

Green crab spider *Diaea* sp. female and male (similar in size)

67

ULOBORIDAE
(FEATHER-LEGGED SPIDERS)

Feather-legged spiders are small to medium-sized, not exeeding 10 mm in body length. They derive their common name from the brush of long setae present on the tibiae of the first pair of legs. They have a calamistrum which is set in a characteristic depression of the metatarsi of the hind legs and the tarsi of the same legs have a ventral row of macrosetae. Uloborids are unusual amongst spiders because they lack venom glands, but use digestive enzymes to kill their prey.

Miagrammopes have only four eyes (the anterior row is missing). They construct a horizontal single-line web, with a central capture area of cribellate silk. This is usually situated between two twigs. In Africa, the genus was previously recorded only from East and South Africa.

Uloborus species have eight eyes in two rows with the posterior row strongly recurved. They weave horizontal orb-webs of cribellate silk, which give them a fuzzy, bluish appearance, and they hang upside-down at the hub. Sometimes there is a linear stabilimentum (silk band) on either side of the spider, making it difficult to detect. They are common in various different habitat types, including around buildings. *Zosis geniculata* is a pantropical species, which can also be expected to occur within The Gambia.

Common feather-leg *Uloborus* sp.

Single-line spider *Miagrammopes* sp.

Zodariidae
(Ant-Eating Spiders)

Zodariids are small to medium-sized, eight-eyed spiders that are diverse in the Afrotropical region, although most of the known biodiversity has been described from South Africa. They are ground-living, wandering spiders and are most easily identified by their long anterior spinnerets (posterior and medians are much reduced) on a distinct base. However, this character is found only in the subfamily Zodariinae. Their colour is usually dark, but the abdomen often has a pattern of chevrons or blotches, which are yellow in the only known Gambian species observed to date.

Many zodariids are good diggers and can be found beneath logs and stones, sometimes in silken retreats. Although seldom seen during the day, unless they are specifically searched for, they can be collected overnight in rather large numbers by using pitfall traps and it is clear that they form an important part of the ground-layer spider fauna, particularly in more arid habitat types, where these spiders appear to be drought resistant.

Little is known about the biology of this family, but many species are thought to be specialist predators of termites or ants, although few species mimic their prey. This certainly seems to be true of the Gambian species pictured below.

Ant-eating spider consuming a large ant

OTHER ARACHNIDS

There are several other groups that share a common ancestry with spiders and have similar morphological features, including two main body parts (sometimes fused or segmented), pedipalps (variously modified) and eight legs. Those known to occur in The Gambia are whipscorpions (order Amblypygi), vinegaroons (order Thelyphonida), hooded tick spiders (order Ricinulei), sun spiders (order Solifugae), laniatores and harvestmen (order Opiliones), ticks and mites (order Acari), scorpions (order Scorpiones) and pseudoscorpions (order Pseudoscorpiones). Additional arachnid orders include Microwhipscorpions (Palpigradi) and schizomids (Schizomida), both of which probably occur in The Gambia, given that they have been found elsewhere in proximate West African countries. However, both are small, cryptozoic animals and are unlikely to be encountered unless specifically searched for. Very little is known about the biodiversity or ecology of these groups within West Africa and especially in The Gambia. Indeed, most of the photographs here represent the first documented records for the country and there will be additional species present that await discovery.

There is still great debate about how these groups are related to one another and the evolutionary tree has still not been resolved. The most recent proposal is shown below.

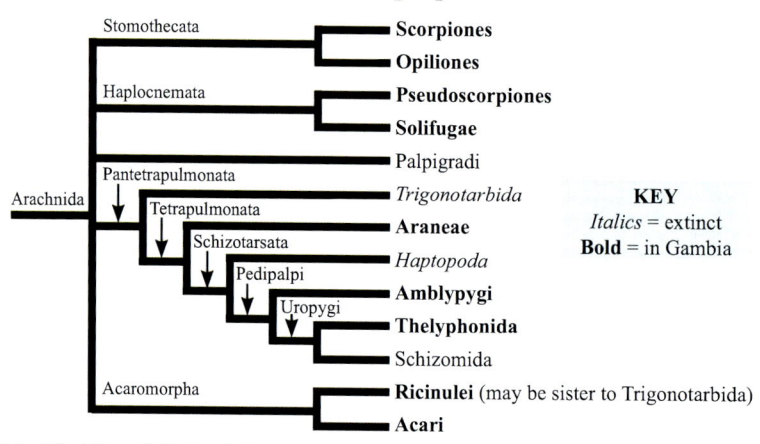

Modified from Schultz, J.W. 2007. A phylogenetic analysis of the arachnid orders based on morphological characters. Zool. J. Linn. Soc. 150: 221–265.

Solifugae
(Camel Spiders)

These unusual arachnids are known by various common names, such as camel spiders, sun scorpions, wind spiders, red Romans, Roman spiders, etc. They live in dry habitats and are usually nocturnal (some are attracted to lights at night), but some species are diurnal and are sometimes seen digging in the sand during the day. They are extremely fast and are voracious predators. Their huge jaws (chelicerae) do not contain venom glands, but can inflict a painful bite and they can quite easily subdue prey larger than themselves, including small vertebrates such as lizards. However, they usually feed on invertebrates, including other arachnids.

The pedipalps are large and give the animal the appearance of having ten legs, but these are used for sensory purposes rather than walking. In addition, they have unusual, stalked mallet-shaped organs on the underside of the last pair of legs, which are also presumed to be sensory, although their exact function is unclear. It is unknown how many species occur in The Gambia and the current classification of this order is considered to be highly unsatisfactory. *Rhagodolus mirandus* (Rhagodidae) has been recorded in The Gambia, with species from the families Daesiidae, Galeodidae and Solpugidae previously recorded from Senegal and Guinea-Bissau.

Sun spider *Zeria keyserlingi* (family Solpugidae, a very fast & voracious predator found in dry areas)

AMBLYPYGI
(TAILLESS WHIPSCORPIONS)

Despite their fearsome appearance, these arachnids (also known as whip spiders) are totally harmless. The first pair of legs are extremely long and antenniforme and these are waved around in a similar manner to insect antennae. They may span up to 44 cm across in particularly large specimens and are used for tactile and olfactory purposes, as well as for communication with other individuals. Species of *Damon* are endemic to sub-Saharan Africa, but only one of the ten or so known species occurs in The Gambia.

They prefer dark, humid habitats and usually live beneath bark or logs in forest leaf litter and move sideways very quickly when disturbed. Their mottled and flat appearance makes them very difficult to see on certain substrates. They sometimes enter houses during the night and can be quite an alarming sight for those who have never seen one before.

The order is divided into two suborders: Euamblypygi, which includes the species recorded from The Gambia and Paleoamblypygi, which is considered very primitive (a 'living fossil') and is represented by the single species *Paracharon caecus* (family Paracharontidae). It is known only from termite nests in Guinea-Bissau, so would not be a surprising find in The Gambia, but would certainly be a noteworthy one.

Whipscorpion *Damon medius* (family Damonidae, first pair of legs extremely long and sensory, harmless)

Thelyphonida & Schizomida (Uropygids)

Thelyphonida (vinegaroons) are highly distinctive creatures that are easily identified by their long, slender whip-like tail. They are not poisonous, but can pinch with their large pedipalps. Little is known about their biology or ecology. They are nocturnal, spending the day beneath stones or logs. They feed on other invertebrates, which they subdue with a spray from the tail. This acts as a solvent, attacking the cuticle of the prey. There is a single species in Africa (two were originally described, but they are now considered to be the same), with a distribution restricted to The Gambia and Senegal. However, it has been suggested that this species may have been introduced from Asia, rather than being an ancient relict.

Historically, Uropygi has included another arachnid group, the Schizomida, which is now given its own ordinal status. These organisms are smaller (less than 8 mm long) than vinegaroons and do not possess such a long, whip-like tail, although they have a short one consisting of three segments. They are usually found in leaf litter or under stones. Although this order is currently not recorded from The Gambia, it probably occurs there, given that species of *Artacarus* and *Schizomus* have been described from Guinea-Bissau, Sierra Leone and Liberia.

Vinegaroon *Etienneus africanus* (family Thelyphonidae, the only African species, restricted to West Africa: The Gambia and Senegal)

SCORPIONES (SCORPIONS)

A much maligned creature, but one that is rarely encountered. Nonetheless, it has been estimated that 5,000 people die worldwide each year from scorpion stings, although most species are harmless and none of the classically dangerous species have been recorded from The Gambia. Only two families are known from West Africa, both of which occur in The Gambia. Five species have been recorded (those below + *Centuroides margaritatus*), although several others have been predicted to occur along the entire West African coast (i.e., Buthidae: *Isometrus, Uroplectes*; Scorpionidae: *Opisthacanthus*).

Pandinus gambiensis
(family Scorpionidae, 17 cm long!)

Hottentotta hottentotta with offspring
(family Buthidae, painful sting)

Butheoloides sp. (family Buthidae, venom toxicity unknown)

Buthus occitanus (family Buthidae, under stones, painful sting)

PSEUDOSCORPIONES (PSEUDOSCORPIONS)

This diverse order of arachnids was previously unrecorded from The Gambia, but they are common in leaf litter and beneath peeling bark. Most species are tiny (less than 5 mm) and so are easily overlooked. They resemble scorpions but lack the long tail and venomous sting. However, most have poison glands in their pincers, which they use to subdue their insect prey. Their prey are detected by means of long, sensory hairs on the claws. Most species do not have eyes.

Pseudoscorpions have silk glands in their chelicerae which they use to produce moulting and brooding chambers. Some species disperse by holding on to the legs or wing cases of flying insects or other animals using their pincers, a process known as phoresy. The offspring are carried on the back of the mother for a short time in a similar manner to scorpions and some other arachnids.

It is not known how many species occur in The Gambia, but they are not uncommon, even in domestic situations. They are often found in bags of rice, where they prey on various pest species. Those families that can be expected to occur within The Gambia include: Feaellidae, Tridenchthoniidae, Chernetidae, Atemnidae and Withiidae, in addition to several others. Five species have been recorded from Senegal, nine from Guinea-Bissau, but more can be expected. One hundred and one are known from the Democratic Republic of Congo!

Afrogarypus sp.
(family Geogarypidae)

ACARI
(MITES & TICKS)

The largest order of arachnids and by far the most important medically (e.g., ticks as vectors of disease in humans and livestock) and economically (e.g., mites as pests). Many are parasites of other animals, such as birds, reptiles and arthropods. They abound in all habitats, including aquatic. Little is known about The Gambian fauna specifically, although important tick genera that occur include: *Argas, Amblyomma, Haemaphysalis, Hyalomma, Ornithodoros, Rhipicephalus*. Other genera recorded include: *Tetranychus, Euschoengastia, Elianella, Sarcopes, Aceria, Tanautarsala* and *Proctophyllodes*.

Velvet ground mite *Dinothrombium tinctorium* (family Trombidiidae)

Millipede (family Odontopygidae) host to numerous mites

Tick *Amblyomma variegatum* (family Ixodidae, a parasite of livestock)

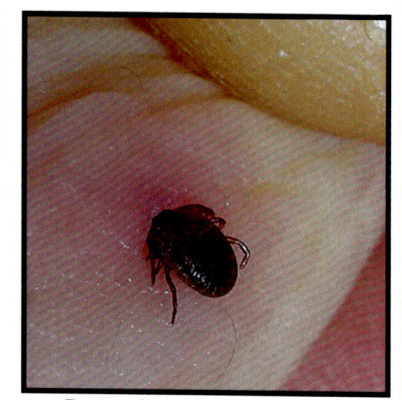

Brown dog tick *Rhipicephalus sanguineus* in the skin of the author

Ricinulei
(Hooded Tick Spiders)

In Africa ricinuleids are restricted to the western tropical region, with records ranging from Guinea-Bissau to southern Cameroon. They are seldom seen and these are the first records from The Gambia, which also represent the northernmost African record of the order. Identification of the species will require microscopic examination, but it is probably *Ricinoides afzelii* or *R. feae*. Adults are grey, juveniles are red.

Many ricinuleid species are known only from a single locality, and some may possess naturally small distributions. This places them at risk of extinction through clearing of primary forest and similar habitats, especially in West Africa. Indeed, it is interesting to note that both localities where the specimens were found are relict fragments of primary forests. Bijilo Forest is the only remaining example of primary rhun palm forest along the coast of The Gambia, whereas Abuko is a relict fragment of primary gallery forest. The Gambia lies within the transition zone at the interface of relatively moist Guinean forest-savannah mosaic south of the river Gambia and drier Sudanian woodlands north of the river. All African ricinuleid species live in moist soils of forest habitats. Thus, it would not be unreasonable to expect The Gambia to represent the northernmost part of the range on the continent.

Hooded tick spider *Ricinoides* sp. grey adult/red nymph

OPILIONES
(HARVESTMEN)

Harvestmen are a diverse group of arachnids, but are easily identified by having both main body regions fused together and the second pair of legs are usually longer than the others. Many species have extremely long and slender legs, but in some species they are relatively short. They are cryptic animals that live in dark, humid conditions, such as beneath logs and stones, in tree holes (or caves) or deep within moist leaf litter. However, some species are adapted to open landscapes, such as grasslands. Many species appear to aggregate and it is often possible to find several or lots of individuals together. They are primarily predators of insects, but some also feed on dead animals or plant juices.

There is barely any non-technical information available on West African Opiliones and they have not previously been recorded from The Gambia. This is possibly in part because they are cryptic and seldom seen, and also because they are harmless and of no consequence as pests of crops. Nonetheless, they are an important part of forest ecosystems and are worthy of further study. It has recently been suggested that Opiliones are more closely related to Scorpiones (and vice versa) than they are to any other arachnid order and both have been placed together in a clade called Stomothecata.

Laniatore (family uncertain)   Harvestman (family Phalangiidae)

Further Reading

AFRAS (African Arachnological Society). http://www.afras.co.za
Azarkina, G.N. 2009. A brief note on Gambian jumping spiders (Araneae, Salticidae). Newsl. Brit. Arachnol. Soc. 114: 7–8.
Belfield, W. 1956. A preliminary check list of the West African scorpions and a key for their identification. J. West Afr. Sci. Assoc. 2: 41–47.
Dippenaar-Schoeman, A. & Jocqué, R. 1997. African Spiders. An Identification Manual. Agricultural Research Council, Pretoria, 392 pp.
Foelix, R.F. 1996. Biology of Spiders (2nd edition). Oxford University Press, New York, 328 pp.
Goyffon, M. 2002. Le scorpionisme en Afrique Sub-Saharienne. Bull. Soc. Pathol. Exot. 95: 191–193.
Harvey, M.S. 2003. Catalogue of the Smaller Arachnid Orders of the World: Amblypygi, Uropygi, Schizomida, Palpigradi, Ricinulei and Solifugae. CSIRO Publishing, 385 pp.
Harvey, M.S. 2007. The smaller arachnid orders: diversity, descriptions and distributions from Linnaeus to the present (1758 to 2007). Zootaxa 1668: 363–380.
Heurtault, J. 1984. Identite d'*Hypoctonus africanus* Hentschel et d'*Hypoctonus clarki* Cooke et Shadab (Arachnides, Uropyges). Revue arachnol. 5: 115–123.
Kovařík, F. 2007. A revision of the genus *Hottentotta* Birula, 1908, with descriptions of four new species (Scorpiones, Buthidae). Euscorpius—Occ. Pubs. Scorpiology 58: 1–107.
Lawrence, R.F. 1969. The Uropygi (Arachnida: Schizomidae) of the Ethiopian region. J. Nat. Hist. 3: 217–260.
Leeming, J. 2003. Scorpions of Southern Africa. Struik, Cape Town, 88 pp.
Legg, G. 1977. Two new ricinuleids from W. Africa (Arachnida: Ricinulei) with a key to the adults of the genus *Ricinoides*. Bull. Brit. Arachnol. Soc. 4: 89–99.
Penney, D. 2008. Possible anti-predator function for the stabilimentum in a West African (Bijilo Forest, The Gambia) *Cyclosa* (Araneae, Araneidae). J. Afrotrop. Zool. 4: 143–146.
Penney, D. 2009. Wildlife of The Gambia: a field guide to common plants and animals. Siri Scientific Press, Manchester, 120 pp.
Penney, D. & Gabriel, R. 2009. Feeding behavior of trunk-living jumping spiders (Salticidae) in a coastal primary forest in The Gambia. J. Arachnol. 37: 113–115.
Penney, D. et al. 2009. First Gambian Ricinulei and the northernmost African record for the order (Arachnida: Ricinulei: Ricinoididae). Zootaxa 2021: 66–68.
Pinto-da-Rocha, P., Machado, G. Giribet, G. 2007. Harvestmen: The Biology of Opiliones. Harvard University Press, 608 pp.
Punzo, F. 1998. The Biology of Camel-Spiders (Arachnida, Solifugae). Springer, 312 pp.
Tuxen, S.L. 1974. The African genus *Ricinoides* (Arachnida, Ricinulei). J. Arachnol. 1: 85–106.
Walker, A.R. et al. 2007. Ticks of Domestic Animals in Africa: a Guide to Identification of Species. Bioscience Reports, Edinburgh, 221 pp.
Weygoldt, P. 2000. Whip Spiders: Their Biology, Morphology and Systematics—chelicerata: Amblypygi. Apollo Books, 163 pp.

Index To Genera Cited

Achaearanea............6
Aceria......................76
Africactenus............28
Afrofilistata..............32
Afrogarypus.............75
Afropisaura..............11
Amblyomma............76
Amyciaea...........65, 66
Anahita....................28
Apochinomma.........27
Araneus..............23, 24
Argas.......................76
Argiope...............22, 25
Argyrodes......41, 63, 64
Ariadna...................59
Artacarus.................73
Badia......................48
Bianor.....................55
Butheoloides...........74
Buthus...............11, 74
Caerostris................24
Castianeira.........11, 27
Centuroides............74
Cheiracanthium......39
Cithaeron................26
Coleosoma..............63
Copa.......................27
Ctenus....................28
Cyclosa.........11, 22, 24
Cyrba......................55
Cyrtophora....23, 63, 64
Gasteracantha.19, 22, 25
Damon...................72
Deinopis.................29
Diaea......................67
Dictyna..................30
Dinothrombium......76
Elianella.................76
Etienneus...............73
Euprosthenopsis....52
Euryopis................63
Euschoengastia.......76
Evarcha........54, 56, 57
Geolycosa........37, 38
Haemaphysalis........76

Hahnia....................34
Hamataliwa........46, 47
Hersilia...................35
Heteropoda........11, 61
Hippasa.............37, 38
Holcolaetis.....11, 54, 56
Hottentotta.............74
Hyalomma..............76
Hyllus.....................56
Ischnothyreus..........45
Isometrus................74
Isoxya................22, 25
Latrodectus........63, 64
Leptopholcus..........50
Leucauge................62
Lycosa...............37, 38
Meleon...................55
Menemerus....11, 54, 55
Messapus................27
Miagrammopes...18, 68
Micropholcus..........50
Minosia..................33
Mizaga..............11, 30
Myrmarachne...54, 57
Mononychellus......11
Muizenbergia.........34
Mysmenella............40
Natta......................57
Neoscona...........23, 24
Nephila............11, 12,
    14, 21, 41, 42, 63, 64
Nesticodes.........63, 64
Nigma....................30
Oecobius................44
Olios......................61
Opisthacanthus.......74
Orchestina.............45
Ornithodoros.........76
Oxyopes...........46, 47
Pandinus...............74
Panonychus............11
Paracharon............72
Parajotus................57
Pararaneus........22, 23
Peucetia...........46, 47
Pehrforsskalia.........50

Perenethis...............52
Philodromus............49
Pirata......................37
Platythomisus.....65, 66
Portia.................54, 55
Proctophyllodes.......76
Prodidomus.............53
Rhagodolus.............71
Rhipicephalus..........76
Ricinoides...............77
Rothus....................52
Sarascelis................48
Sarcoptes................76
Schizomus..............73
Scytodes.................58
Selenops.................60
Smeringopus..........50
Speocera................43
Steatoda.................63
Stegodyphus...........31
Stenaelurillus.....11, 56
Tanautarsala..........76
Tetragnatha...........62
Tetranychus...........11
Thanatus................49
Theridion...........63, 64
Thomisops..............67
Thomisus..........65, 67
Thyene...................57
Tibellus..................49
Tmarus..............65, 66
Trachelas................11
Uloporus................68
Uroplectes..............74
Xysticus.............65, 66
Zelotes...................33
Zeria......................71
Zimiris...................53
Zosis......................68